GREENHOUSE and INDOOR PLANTS

Also available in the Pan Library of Gardening:
Roses and their cultivation
Garden Shrubs and Trees
Fruit and Vegetables

Conditions of Sale
This book shall not, by way of trade or otherwise, be lent, re-sold, hired out or otherwise circulated without the publisher's prior consent in any form of binding or cover other than that in which it is published and without a similar condition including this condition being imposed on the subsequent purchaser. The book is published at a net price, and is supplied subject to the Publishers Association Standard Conditions of Sale registered under the Restrictive Trade Practices Act, 1956.

This book may not be sold in the
United States of America or Canada.

Pan Library of Gardening

GREENHOUSE and INDOOR PLANTS

Edited by Peter Hunt
Introduced by Thomas Rochford, V.M.H.

 Pan Books Ltd: London

Published 1972 by
Pan Books Ltd.
33 Tothill Street, London SW1.

ISBN 0 330 02918 5

© Marshall Cavendish Ltd., 1969, 1972

Printed by Proost, Turnhout, Belgium.

The material contained in this book is based on material first published in 'The Marshall Cavendish Encyclopedia of Gardening', edited by Peter Hunt.

Contents

Chapter		Page
1	**Cultivation**	9
2	**Suitable house plants**	17
3	**Greenhouse management**	27
4	**Electricity in the greenhouse**	39
5	**What to grow in your greenhouse**	45
6	**Pot plants**	53
7	**Propagation**	63
	Index	71

Glossary

Acaricide A chemical which will kill mites, usually red spider mites. The term is applied to manufactured dusts and sprays which are used against mites.

Aroid Plants belonging to the botanical family *Araceae*, e.g. the arum lily.

Bract A modified leaf at the base of the flower stalk, or forming part of the flower-head.

Cultivar A plant which has originated under cultivation and is sufficiently different from others of its kind to be given a separate varietal name.

Epiphyte (epiphytic) A plant which grows upon another plant without actually being parasitic.

Eye A small, vital part of a plant.

Fungicide A chemical substance which is used in the control of those diseases of plants which are caused by fungus parasites.

Palmate Shaped like an open hand, referring to leaves with spreading lobes or finger-like divisions.

Pendulous Drooping.

Periclinal chimaera Unusual plants are sometimes produced when tissues of two separate kinds grow over each other. Such plants are known as chimaeras. Most plants which have leaves with a colourless or nearly colourless edge are chimaeras, of the type known as periclinal chaemaeras.

Pollinate To fertilize with pollen.

Rhizome An underground stem or creeping rootstock.

Terrestrial Plants growing in soil as opposed to aquatics and epiphytes.

Introduction by Thomas Rochford, V.M.H.

House plants have become a necessary part of our daily lives. Like the rest of the decorations in our homes the house plant takes up an important place with our pictures, china and furniture as a part of the permanent overall effect. Their purpose must not be confused with that of cut flowers; these are lovely, and we all want them in our homes, but they have a different use. They can only be temporary, and are probably most often used for special occasions. House plants have a different quality, they can become part of the family, and live and grow with the rest of the household.

This important addition to the home appeared in people's houses and flats in a new form about twenty years ago, when architects began to give us new designs which created conditions and surroundings to suit the many varieties of handsome foliage plants. Better light, air conditions and heating all played an important part.

In Victorian times, the interest was there, but conditions were such that only the very tough plants such as palms, aspidistras and some ferns could survive for any length of time. Now we have a choice from hundreds of beautiful plants, and this in its turn has caused increasing interest. Today nearly everyone can grow the philodendrons, hederas (ivies), ficus, cissus, peperomias, marantas, all the bromeliads, and a host of others. The choice is endless and there are plants of every shape and size and of all colours to suit every setting. It is interesting to realise how much foliage plants can come to mean to us in our own surroundings; often as much care is taken with them as we give to the family cats and dogs. Our children grow up with them and the next generation with think of them as necessary to their own homes. Where people are living in tower blocks of flats or homes with no gardens, they can now have something green and living in their rooms.

This is one of the reasons why demand has escalated so rapidly and has swept through the United States and Europe. I do not think that there is any question of house plants being a mere craze, they have a real purpose in every home, and I am sure they will always be wanted.

The term 'pot plant' covers all plants that are grown in pots, but not all of them should be called house plants. Flowering pot plants should not be put into the house plant category. They are grown for their flowers and not their foliage, so that when the flowers have gone the leaves tend to be uninteresting, and the plant is usually removed from the room and is often returned to the greenhouse. Flowering pot plants will last longer than cut flowers but they cannot give the long lasting attractions of a house plant.

There are of course some plants that have flowers and also fine leaf markings, and these are classed as flowering house plants. A good example is the aphelandra which has beautifully marked leaves, and when the flower head has finished, it can be removed, leaving a plant with fine good looking foliage.

The qualities we look for in house plants at Rochfords, in order that we can offer them to the public, are those which make us confident of their ability to thrive in shady conditions. However light a room may appear to be, it is comparatively shady if you think of conditions out of doors. Near the windows the light will be fairly clear, but moving back into the room the diffused light will be much the same as it is where they grow in their native forest conditions.

Many of the house plants we know grow naturally at the bases of trees in forests all over the world. We select plants that have a robust constitution capable of coping with fluctuations in temperature and fumes from gas and oil.

The all important item in maintaining the health of house plants and greenhouse plants lies in correct watering. Generally speaking the soil should be on the moist side, but do not let the plants suffer extremes of being either very wet or very dry. However, a very dry plant will have a better chance of survival than one which has been drenched. There is an exception in the popular sansevieria which should always be kept on the dry side.

All these points have been admirably dealt with in the relevant chapters of this book. I am sure that the book will help every house plant grower and greenhouse gardener to cultivate his plants in the most successful way, and I hope that many readers will go on to find other varieties of house plants and greenhouse plants that are new to them and will add them to their collection.

This is, after all, much the same way as our collection was built up at the House of Rochford. About 23 years ago Rochfords had some twenty to thirty varieties, and now we have between 300 and 400, grown on 42 acres. We are in contact with all parts of the world in the search for new species and cultivars. Many millions will be grown next year and for many years ahead, because we know that interest and demand are rapidly increasing all the time. The introduction of the house plant has brought pleasure to millions of people.

Cultivation

There are two kinds of plant that we bring into our houses. The more spectacular are the flowering plants, cyclamen, azaleas or African violets. Unfortunately their season of attractiveness is limited. All too soon the flowers will fade and the plants then have little attraction. If you have a greenhouse, you can keep the plant going and prepare it for another season, but we usually do not feel inclined to keep it in the house; certainly not in a conspicuous position. The other kind of plant is less spectacular; its beauty is centred in its foliage rather than in its flowers, but it has the advantage that, provided you treat it properly, it will grow permanently in your rooms and increase in size and effect from year to year. These plants, grown for permanent effect, are commonly known as house plants.

Most of us do not live in glass houses; therefore the plants we can grow in our rooms are limited in number. Even a room that appears well lit to us, will seem shady to a plant and, as a result, the majority of houseplants are those that can tolerate shade. If a plant is to be permanently attractive, it should be evergreen. We can visualise exceptions, such as the bonsai dwarf trees, where the outline of the tree is attractive even when no leaves are visible, but there are not many of these exceptions to the demand for evergreen plants. Again most of us live in rooms of only moderate size and we require plants of moderate dimensions. We also do not want them to grow too quickly. Re-potting is a tedious operation for those who live in flats or in houses without gardens and we do not want to have to undertake it more than once a year at the most. Although plain green leaves are attractive enough, particularly if they have interesting shapes, leaves that contain some colour are usually regarded as more attractive. Colour in leaves occurs in two forms. Some leaves are naturally coloured; for instance those of Rex begonias and *Cordyline terminalis*, but there are other plants which produce coloured forms of their normally green leaves. These are described as variegated and the variegation may be due to a number of causes, from a virus infection to a periclinal chimaera.

Whatever the cause, the result is that some part of the leaf lacks chlorophyll, the substance that makes leaves green. If the chlorophyll is completely lacking, the area appears white, while, if there is very little present, the area appears golden or yellow. Although some plants appear naturally variegated, nearly any plant can occasionally produce a variegated form. Variegation is an exceptional occurrence and the plant can be perpetuated only by vegetative propagation, normally by rooting cuttings or layers. Seeds are very unlikely to transmit the variegation. By no means all people find variegated leaves attractive, but very many do and, as a result, many plants are popular because of their variegated leaves that would otherwise be little regarded. The popular variegated forms of *Tradescantia fluviatilis* and *Chlorophytum capense* may be cited as examples. With half or more of the chlorophyll lacking, the leaves of variegated plants can only do half the work of normal green leaves and so variegated plants tend to grow more slowly than the unvariegated forms. This is not unexpected; a more surprising result of variegation, though it is not always the case, is that the plants are often more tender than the normal forms. There seems to be no very obvious explanation for this.

We can now summarise the qualities that we require for a house plant. It must be compact in habit of growth, tolerant of shade and evergreen. The leaves should be attractive, either by reason of their shape or their colour: if we can have attractive flowers as well, so much the better, but with the emphasis to be laid on the permanence of the attraction, agreeable flowers are obviously a bonus. In fact the combination of handsome flowers and handsome leaves is somewhat rare in any branch of gardening. Among the house plants many of the bromeliads provide an exception to this rule, but even with these plants the showiest part of the inflorescence is due to the coloured bracts that surround the flowers, and these are really modified leaves.

Temperate climates produce few plants with the characteristics that we require. The various ivies form an important exception to this statement, but, even so, the great majority of house plants come from the tropics. Plants are infinitely adaptable, as a general rule (there are, of course, exceptions and these are generally regarded as 'difficult' plants) and most tropical plants will adapt themselves to temperate conditions and even to the fluctuating lengths of daylight, which, even more than the alteration in temperature, mark the chief difference between tropical and temperate climates.

There are certain temperatures, vary-

A group of flowering house plants all of which add colour indoors in winter.

ing from plant to plant, below which plant growth ceases. The plant may survive perfectly well, but it will neither produce fresh roots nor fresh leaves, until the temperature is raised. It is, obviously, more difficult to produce high temperatures when the outside temperature is very low and so it is most convenient to make our winter the equivalent of the tropical plant's dry season. The dry season in the tropics is generally very hot, but, owing to the lack of water, the plant makes no growth and stays in a dormant condition. This is one of the reasons why all house plant growers are recommended

Evergreen plants grown for indoor decoration provide interest all year round. Many have attractive leaves, like this Fittonia argyroneura.

to keep their plants as dry as possible during the winter. How dry you can keep them, will depend on the type of plant you are growing and on how warm you keep your rooms.

The type of heating that you use and the temperatures you maintain in your various rooms during the winter will affect the types of plant you can grow. Some plants, notably begonias, are very intolerant of gas fumes, so that if your rooms are heated by gas, you will not be able to grow begonias satisfactorily. If you have really warm rooms, maintaining, perhaps, an average temperature of 70°F (21°C), they will be far too warm for such plants as ivy or × *Fatshedera lizei*. With high temperatures such as these, the plants will continue growing during the winter and more water will be required. The winter growth may not be very ornamental, as the lack of light will prevent the formation of good sized leaves.

Many house plants are 'stopped' in the spring: that is to say that the tips of the various shoots are nipped out, so as to induce the formation of secondary shoots that will give the plant a nice bushy appearance, and where this is done the weak winter growth can be removed. However, there are plants, such as most of the ficus, that are not stopped and, where these are concerned, it might be better to move them to cooler positions in the winter. However most of us, alas, cannot afford these high temperatures and it is more a question of keeping the room warm enough for our plants and ourselves. In any natural climate the highest temperatures are around midday, but many sitting rooms are kept cold during the day, when people are out at work, warm in the evening, when everyone is at home, but cool off during the late evening and the early morning after people have gone to bed.

Such a contradiction of natural rhythm is sufficient to disturb any plant and it is easy to see that keeping plants in good condition in the winter is less simple in the house than in a greenhouse. If you have some system of regular central heating, the problem is comparatively simple, but for rooms with only sporadic warmth, the matter is less straightforward. However, there are house plants to suit all conditions. It is as well to know what the average temperature of your room is during the winter, otherwise the problem can be resolved only by a system of trial and error, during which you might well lose the plants that you most prize. It is fairly safe to say that no plant will tolerate the conditions that are to be found on a mantelpiece above a coal fire. The atmosphere is far too dry and the alternations of cold and roasting heat are too much for all plants, except the toughest succulents. Even if the temperature is equable, plants that are put too near the window risk being chilled, or even frosted, when the weather is very cold and they should be moved further into the room during these periods.

Even when the temperature is satisfactory, the dry atmosphere that we like in our rooms is not beneficial to plants. This, however, can easily be overcome, by placing the pots in a larger container and filling this container with some moisture-retentive material. Peat is most frequently used, but moss or mica powder does equally well. Some people get perfectly satisfactory results with damp newspaper, which is topped with moss to look more elegant. By these means we can maintain a moist atmosphere in the immediate surroundings of the plant without either affecting the atmosphere of our rooms or the correct state of moisture of the soil ball. With this we come to the most crucial matter in the successful cultivation of house plants.

More house plants are killed by overwatering than by any other cause. Like human beings, plants cannot live without water, but, again like human beings, they can be drowned. However, this analogy cannot be pressed too far. Human beings need water at regular intervals, but plants need water most when they are making growth. This is usually during the spring and summer. There is a correlation between the growth of the aerial portion of the plant, the portion we can see, and the growth of the roots which we can't see. If the plant is making new leaves and stems, we can be fairly sure that it is also making new roots. Unfortunately, the root growth is liable to precede the production of new leaves and so these latter may be prevented from developing if the soil is too dry at the appropriate period. On the other hand, if the soil is too wet, the roots cannot breathe and, far from developing, are liable to rot and, unless this process can be stopped immediately, the plant itself will succumb. We can guard against this to some extent by purely physical means. If we have an open soil mixture that drains

rapidly and well, the risk of the soil becoming sodden and sour is reduced, although not, of course, obviated altogether. When to give water is only satisfactorily learned by experience, but the following hard and fast rules are generally acceptable.

1 When water is applied, it should be in sufficient quantity to moisten the whole of the soil ball. The water should be at room temperature. Rain water is preferable but not essential.

2 The soil should be allowed to dry out between waterings. This is not too easy to interpret. We do not want the soil to become dust dry, but on the other hand, we want to avoid saturation. A useful rule of thumb with clay pots and soil mixtures, is to strike the side of the pot with your knuckle. If the resultant sound is dull, watering is not needed, but if it is a ringing sound, water should be applied. With peat mixtures the weight of the pot is a more reliable indication. If it feels light, water is wanted, but not if it feels heavy. The same applies, but to a lesser extent, with soil mixtures in plastic pots. These are much the most difficult to gauge.

3 During cold weather plants make little or no growth and so require little water. Growth is also slowed down when there is little light. It is safe, therefore, during the winter to keep all watering down to a minimum, even though the room may be kept at quite a high temperature. Naturally plants in warm rooms will require more than those in cool

Hedera helix 'Chicago', a pointed-leaved Ivy with (front) **Chlorophytum comosum, sometimes known as the Spider Plant.**

surroundings.

4 From about mid-April it is probable that growth will start and so more water may be required. Be cautious, nevertheless, until you see new leaves appearing. It is possible to knock the plant out of its pot to see if new roots (characterised by their white tips) are forming and to replace the soil ball without disturbance. When growth is vigorous water will be needed more frequently. 'Stopping' checks growth temporarily and watering should be on a reduced scale until a resumption of growth is seen.

5 By the end of August it is advisable to discourage much further growth and encourage the plant to ripen its new growth. This is done by keeping the plant as dry as possible.

6 The type of leaf will give some indication of the plant's requirements. Plants with thick leaves or with succulent leaves (such as the large-leaved ficus and sansevieria) can tolerate longer periods without water than thin-leaved plants. These latter will probably wilt when they become too dry and they should be watered at once. The thicker leaved plants will not wilt and so should be inspected frequently. Drought, unless acute, will not kill them but may cause subsequent leaf drop.

If leaves turn yellow and fall off, it usually indicates over-watering. However some plants, such as *Ficus benjamina*, will naturally shed their year-old leaves in the autumn and most plants shed a few leaves in the course of the year. Excessive defoliation is almost certainly due to incorrect watering; although it can be caused by under-watering as well as by over watering. If the plant becomes unsteady in the pot, this is generally due to root-rot caused by excessive water and is very difficult to arrest. Some leaves will wilt in the summer if they are in direct hot sunlight. If the soil appears to be satisfactorily moist, a syringeing of the leaves with water will generally restore them to their normal turgidity, and in any case, they will resume their normal appearance as soon as the sunlight goes.

Once the question of watering has been mastered, there are few other problems. Rooms are very dusty which spoils the appearance of the leaves of house plants and also prevents them from functioning properly. It is advisable, therefore, to clean the leaves every two or four weeks. This is best done with cotton wool and tepid water and the leaves should be sponged on both sides. New leaves are soft and easily damaged and should be left until they are older. Some people use milk, or oil, or flat beer to give the leaves a more glossy appearance, but these mixtures do not do the leaves any good.

Maranta mackoyana must be protected from sunlight to keep its leaf colour.

During the summer, when growth is most vigorous, the plants may be fed. A liquid feed is most easily applied and should be given according to the instructions on the bottle. Little and often is invariably better than doses in excess of those recommended. Unless the plant is really well-rooted, feeding should not be applied and is not necessary for plants that have been repotted. Repotting is done in the early summer. For the majority of house plants the John Innes potting compost No 2 is the best. Plants are usually potted on into a pot one size larger. Plants from 5-inch pots are put into 6-inch pots and so on. The only exception is that the 4-inch pot is very rarely used and plants are moved from 3-inch to 5-inch pots. Plants with very thin roots such as begonias and peperomias do better in a mixture of 2 parts of leafmould to 1 part of sharp sand, while epiphytes, such as the bromeliads, are usually given a mixture of peat, leafmould and sharp sand. However, it is only rarely that epiphytes require any potting on, as they use the soil as an anchorage only. After being potted on the plants should be kept on the dry side until the roots have penetrated the new soil. It is best to move plants from 2-inch pots to 5-inch pots after a year, as the 3-inch pots dry out so quickly, but after that most house plants will need repotting only every other year. The second year the plant will need feeding.

The epiphytic bromeliads (aechmea, neoregelia, nidularium, guzmania, tillandsia and vriesia) need rather different treatment from most house plants. They have a rosette of strap-shaped leaves which form the so-called 'vase'. This must be kept full of water, preferably rain water. The mixture in which they are potted may be kept moist, but this is of minor importance, as the roots serve little purpose except anchorage and it is from the leaves that nutriment is absorbed. During the summer the merest trace of liquid feed may be added to the water in the vase, but this must be done with great discretion.

Most house plants are easily propagated if you have a greenhouse, although some, such as the large-leaved ficus, cordylines and dracaenas, need a good deal of heat to get them to root. There are a few that can easily be propagated in the home. The various tradescantias and zebrinas will root easily in water and so will *Cissus antarctica* and *Rhoicissus rhomboidea.* Shoots of succulents, such as *Sedum sieboldii* and aichryson will root easily, either in ordinary soil or in a mixture of equal

A collection of Ferns in a bark container, with Helxine added in the front.

Peperomia magnoliaefolia needs far less water than most house plants.

parts of peat and sharp sand, which is an ideal mixture for most cuttings. Shoots of the various ivies, taken when they are half ripe, that is neither too young nor too woody, will root easily although rather slowly. The peperomias, with single leaves rising from the base, can be rooted from leaf stem cuttings. The leaf with its stem is pulled off and inserted in the peat and sand mixture, when a new plant will form at the base of the leaf stalk. Sansevierias will produce new leaves on rhizomes, but they do not root until a year has elapsed and should not, therefore, be severed from the main plant before this time. If, however, you can find the new rhizome without disturbing the plant and cut half-way through it, it will hasten the formation of roots at the base of the new leaf. Many of the climbing aroids produce aerial roots and these can be induced to develop in soil. These climbing aroids (philodendron, syngonium and scindapsus), will grow more luxuriantly if they are given a cylinder of wire stuffed with moss up which they can grow. However, the moss must be kept damp and this is not easy in the home. They can also be trained on blocks of cork bark. (See also chapter 7.)

little trouble with either insect or fungus pests on your house plants. The most troublesome is liable to be red spider mite. This is a tiny mite which congregates in large numbers on the undersides of leaves, causing discoloration. The mites are not easy to see with the naked eye, but affected plants have a characteristic rough feel: This pest is usually associated with very dry conditions and these are liable to occur in rooms during the summer. If the leaves are regularly sponged, it is improbable that any serious infestation will occur. If it does, it is necessary to immerse the plant in water containing a suitable acaricide or in a white oil emulsion. Aphids may be found occasionally, clustering around the growing points, but aerosol sprays are available to control them. Some of the larger-leaved ficus may have scale insects on the underside of the leaves. If the plant is inspected regularly, there is little risk of this becoming serious. The scale insects may be removed with a matchstick and then destroyed.

Suitable house plants

The following lists are a selection only of the more popular house plants, arranged according to their winter heat requirements. There are many more plants that you can grow if you are interested.

Plants for unheated but frost-proof rooms
Aichryson × domesticum variegatum This is a small variegated succulent, a hybrid of garden origin, with long, oval leaves, almost entirely ivory-white when young, later turning olive-green with an ivory margin.
Aspidistra lurida This, with its broad, dark green leaves, was once the most popular house plant of all, and it is once again returning to popularity, particularly the variegated forms.
Billbergia nutans A terrestrial bromeliad with bluebell-like flowers, but rather uninteresting leaves.
Chlorophytum comosum This is an attractive variegated plant, with long, narrow leaves, which forms new plants at the ends of the flower stems. It is sometimes called, inaccurately, *C. capense*.
Cissus antarctica A climbing plant that will grow anywhere except in very deep shade.
Cordyline indivisa and *Dracaena parrii* These are palm-like plants with many long narrow leaves. They should be given well-lit positions.
Cyperus diffusus The umbrella grass is

1 Symbol of Victoriana, Aspidistra lurida, with long shining leaves does well in an unheated room.
2 Cyperus diffusus, the Umbrella Grass, an easily grown plant, needs ample water particularly in hot weather.

very easy to grow, but rather dull on its own.

x *Fatshedera lizei* This is an intergeneric or bigeneric hybrid between *Fatsia* and *Hedera*. It makes a tall leggy plant, with dark green, five-lobed leaves. The variety *variegata*, has leaves with cream margins.

Fatsia japonica The castor oil plant. The variegated form is the most attractive.

Hedera helix There are numerous cultivars of the common ivy. Some are small-leaved, self-branching plants that do not require stopping, as they form sideshoots naturally. They like reasonably light situations and are slightly affected by gas fumes. The most popular cultivars are 'Chicago', 'Pittsburgh', *sagittaefolia*, 'Heisse' and 'Little Eva'. There are other cultivars that are not self-branching and need stopping two or three times in the spring and summer. The most popular are 'Glacier' and 'Golden Jubilee'. The latter, a most attractive plant, needs as bright a light as you can give it. The climbing, larger-leaved, cultivars, *marmorata* and *maculata aurea*, do not branch much, but any sickly winter growth should be removed in the spring. The forms of *Hedera canariensis, foliis variegatis* and 'Golden Leaf', do not break after stopping and can be grown in warmer situations than the forms of *H. helix*.

Pittosporum eugenioides This is a pleasant shrub with black stems and attractively shaped leaves. There is also a good variegated form.

Rhoicissus rhomboidea A vigorous vine-like plant with leaves composed of three leaflets. The best cultivar is 'Jubilee'.

Saxifraga stolonifera The well-known

The Canary Island Ivy, Hedera canariensis foliis variegatis, has colourful variegated leaves.

A shining-leaved climber for a cool room, Rhoicissus rhomboidea is a native of Natal.

Cissus antarctica, the Kangaroo Vine, is very sensitive to gas fumes. It is a native of Australia.

'mother of thousands', with marbled leaves and flowers like those of 'London pride'. There is a variegated form, *tricolor*, which is attractive but difficult to keep in good condition. It probably requires more warmth.

Sedum sieboldii variegatum This has round, blue-green leaves with a yellow centre and heads of small pink flowers in the autumn. It is not truly evergreen and dies down each autumn, to reappear in February.

Stenocarpus sinuatus This is an attractive shrub with oak-like leaves, reddish when they first appear.

Tradescantia fluviatilis and *Zebrina pendula* Both these are known as the wandering Jew, and are very easy to grow and propagate. Only the variegated forms of the trandescantia are worth growing and any unvariegated growths should be removed immediately, when detected, otherwise the plant will revert to the ordinary green form.

Plants for rooms with an average winter temperature of 50°F (10°C)

Begonia rex and *most other Begonias* These like a somewhat shady position and are very susceptible to gas fumes. The pots should be plunged in moist peat to provide a humid atmosphere. The soil should not be over-watered.

Brassaia actinophylla (syn. *Schefflera actinophylla*) This is a vigorous shrub with glossy leaves, divided into five leaflets, eventually up to 1 foot across.

Cissus sicyoides This is a not very vigorous vine with palmate leaves on crimson leaf stalks. It dislikes gas fumes.

Citrus mitis The calamondin is a small orange with fragrant white flowers and

The young leaves of Stenocarpus sinuatus are reddish at first.

small, rather tart, orange fruits. It is best stood outside in full sun during the summer. This ripens the wood and encourages the formation of flower buds, but turns the leaves a rather unhealthy colour. They will usually revert to a pleasant glossy green after being brought indoors again at the beginning of September.

Ficus benjamina An attractive small tree with a slightly pendulous habit, this sheds, each autumn, a quantity of leaves, but as it has previously produced plenty of foliage the plant is always pleasant to look at.

Ficus elastica There are many forms of the India-rubber tree available; the best unvariegated form is *decora*. The variegated forms (*doescheri, schryveriana, tricolor*) need more light.

1 The Rubber Plant, Ficus elastica decora, lives happily in stuffy rooms.
2 The wavy-leaved form of the Indoor Fig is Ficus lyrata.
3 A variegated-leaved form of the Indoor Fig, Ficus schryveriana.
4 One of the Philodendrons that does not climb is P. bipinnatifidum. The leaves may be up to 2ft long.

Monstera pertusa This is usually called *M. deliciosa borsigiana;* it is similar in habit to the climbing philodendrons, but has large leaves that are perforated with circular holes and the margins may be fringed. It needs a good deal of light, as, if it is too heavily shaded, the leaves will not develop the characteristic attractive perforations.

Peperomia magnoliaefolia, P. obtusifolia, and *P. glabella* The peperomias have a very small root system and very rarely require potting on. They will grow in

5 The Rugby Football Plant, or Water Melon Plant, Peperomia sandersii.
6 The arrow-shaped leaves of Philodendron hastatum need frequent cleaning.

1 One of the climbing forms of Philodendron, P. scandens.
2 A young plant of the Mexican Breadfruit Plant, Monstera delicosa borsigiana. It is also sometimes called the Swiss Cheese Plant.

shady situations and need far less water than most house plants.

Philodendron bipinnatifidum This is not one of the climbing philodendrons; it has very large, much divided leaves. Each leaf may be up to 2 feet long and 1½ feet across.

Philodendron scandens, P. erubescens 'Burgundy', P. elegans Climbing plants of the arum family, these have aerial roots rising from the stem at the base of each leaf-stalk. They will grow best if given cylinders of moss on which they can climb, but are satisfactory without this.

Platycerium bifurcatum The elk's horn or stag's horn fern, does well fastened to a block of cork wood. If this is done the plant must be immersed in a bucket of water from time to time, as ordinary watering is ineffective. The plant will grow in a pot, but is less effective when grown in this way.

Sansevieria trifasciata laurentii This has tall mottled leaves with cream margins and is a popular house plant. It should be kept quite dry during the winter, and watered moderately only at other times.

Scindapsus aureus This resembles a climbing *Philodendron*, but has golden variegated leaves. It needs ample light.

Syngonium Another climbing aroid, with leaves shaped like a goose's foot, this is a less vigorous climber than the philodendrons.

Tetrastigma voinerianum A very vigorous vine with leaves composed of five leaflets, that may be 10 inches across, this plant needs good light and plenty of room, as it may grow very tall.

Plants for rooms with winter temperature 55–60°F (13–16°C)
There are many more philodendrons, both terrestrial and climbing. The most attractive is *P. melanochryson*. This has heart-shaped, dark velvety-green leaves, purple-pink on the undersides. It is the juvenile form of the large-leaved *P. andreanum*.

Anthurium scherzerianum This has bright scarlet, waxy, 'painter's palette' flowers. It needs to be grown in a mixture of sphagnum moss and leafmould, which is aerated by broken crocks. It likes a moist atmosphere and must not be allowed to dry out.

Aphelandra squarrosa This is a handsome and popular plant with lance-shaped dark green leaves, with prominent ivory veins and pyramids of yellow flowers. The best cultivars are 'Brockfeld' and 'Silver Beauty'. The variety *louisae* is the one that is most generally grown.

Bromeliads Although the various bromeliads are more tolerant of cold than their exotic appearance would suggest, they

---

3 Sansevieria trifasciata laurentii, or Mother-in-law's Tongue.
4 The deeply cut leaves of Philodendron elegans are an attractive addition to any collection of indoor plants.
5 A young plant of Tetrastigma voineriana.

will do best in this section. They will, however, survive at lower temperatures.
Codiaeum variegatum The 'crotons' have spectacularly coloured leaves in a multitude of different shapes and colours. They will not tolerate draughts, which cause leaf-drop and they need a well-lit position. They should not be stopped until they become too leggy, which means that no stopping should be done for two or three years, and then it is a question of cutting-back, rather than stopping. The sap of the plants is milky and flows freely, so, when plants are cut back, it is useful to have some cotton-wool handy to clean them up.
Cordyline terminalis These are the broad-leaved 'dracaenas' with colours of crimson and red in the young leaves. They are spectacular and not difficult to

1 Sansevieria trifasciata laurentii, and Hedera 'Heiss' in the foreground.
2 Scindapsus aureus, the Devil's Ivy.
3 Dracaena godseffiana has opposite leaves heavily spotted with cream and likes a warm room with plenty of light.
4 The Striped Dracaena produces whorls of green leaves, edged and heavily marked with cream.
5 In older plants of Dracaena marginata, the leaves are edged and veined red.
6 A poisonous plant Dieffenbachia amoena, justifiably called Dumb Cane.
7 Dracaena massangeana, a spreading plant with variegated leaves.

grow, except that they are liable to be infected by red spider mites.
Dieffenbachia amoena The dumb-canes have large, oblong-oval leaves, marbled with cream or yellow. The plants are spectacular, but very poisonous, and they are not recommended where there are children.
Dizygotheca elegantissima (syn. *Aralia elegantissima*) This plant has palmate, spidery leaves when young, although in mature specimens the leaves are considerably broader. Again red spider mites are the enemy.
Dracaena fragrans, D. deremensis, D. goldieana These are very beautiful shrubs with broad leaves, striped with various shades of silver, cream and yellow. The first named has the broadest leaves. They should never be stopped.
Ficus lyrata and *F. nekbudu* These are both handsome, large-leaved ficus. The leaves of the first are shaped like the body of a violin, the second has large, oblong-oval leaves with prominent cream veins. They should be examined frequently for infestation by scale insects. These are

usually on the underside of the leaves and may be removed with a match-stick. Neither plant should be over-watered in winter.

Maranta and *Calathea* These are plants with remarkably attractive leaves, that must be shielded from direct sunlight. They require a moister atmosphere than most houses can provide and the pots in which they are grown should, therefore, be put in bowls of moist peat or something similar.

Peperomia argyreia, P. caperata, P. hederaefolia Peperomias have single leaves rising from a common rootstock. *P. argyreia (P. sandersi)* is the most handsome with its rugby-football-shaped leaves of cream and grey. All these peperomias should be kept fairly dry, but given a moist atmosphere surrounding them.

Spathiphyllum wallisii This is grown for its attractive, white, arum-like flowers, which may appear in the autumn as well as in the spring. *Spathiphyllum* 'Mauna Loa' has much larger flowers, but they are not produced so freely. The soil should not be allowed to dry out in winter.

1 A plant grown for its white Arum-like flowers in winter, is Spathiphyllum.
2 Beloperone guttata, the Shrimp plant, does well in a warm room.

Greenhouse management

Few gardeners do not at some time or another have an urge to grow plants in a greenhouse. It extends the range of plants that can be grown so that those too tender for our winter climate can be cultivated and it is a branch of gardening that can be enjoyed whatever the weather.

Before purchasing a greenhouse you should decide what plants you wish to grow and choose a design which is suitable for them. For instance, if you wish to grow mainly tomatoes in the summer and lettuce in the winter you could well choose a greenhouse glazed almost to the ground as both these crops need maximum light and they benefit from the extra glazing. If you are mainly interested in growing pot plants the greenhouse could have partly glazed side walls, the lower half being of wood or brick with staging for the plants level with the side walls. It is also possible to have a greenhouse glazed to the ground on one side and with a low wall on the other side to enable plants to be grown in a border on the glazed side whilst pot plants of various kinds can be grown on the staging on the other side (see chapter 9).

Heating Before a heating system is installed you should give careful thought not only to its initial cost but also to how much it will cost each year to maintain the desired minimum winter temperature. As a rough guide, to heat a greenhouse to a minimum temperature of 50°F (10°C) will cost twice as much as one heated to 45°F (7°C); three times as much for a temperature of 55°F (13°C) and four times as much for a temperature of 60°F (16°C).

Although an unheated frame or greenhouse can provide much interest and enjoyment, the addition of some form of heating will not only extend the range of plants which can be grown but will add considerably to the pleasure you can obtain from your greenhouse.

With so many different forms of heating appliance available, it is often a little difficult for the amateur to make an easy selection or decision. The three main types of heating are electricity, solid fuel and oil. Each has its own particular merits and drawbacks. The final decision as to the best type to use can be simplified if certain points are carefully considered. The first must be the amount of money which is available, not only for the initial purchase but the running of the apparatus afterwards.

Oil The cheapest to purchase and maintain is the paraffin oil burner. Many different types are available, from the very small model which could be used for heating a frame, to the large types which are quite capable of keeping the temperature inside big greenhouses (20 feet × 10 feet) well above freezing point. Whilst there are some very good models available, there are some which are very cheap and poorly made. It is very important that the purchase of an oil heater is made only from a specialist firm who use good quality metal and provide a suitable burner or wick. The best type of heater incorporates a blue-flame burner which, if properly trimmed and used with a high grade oil, should burn without causing fume damage to plants inside the greenhouse. There is a range of excellent heaters manufactured from solid, hand-rolled copper. These should provide many years of faithful service and will not corrode.

Variation in oil heater design is to be found in the way in which the heat from the burner is utilised. Some models are quite simply metal boxes which collect the heat in their top compartment, then releasing the heat through small holes or vents in the sides. They are very efficient heaters and quickly warm up the interior of the greenhouse.

Many designs make use of tubes or radiators which vary in length according to the capacity of the burners. These tubes also have holes in them through which the hot air is released into the greenhouse. The larger and more expensive types have several rows of pipes some of which contain water with the uppermost pipe acting as a flue or chimney. This is particularly suited to the larger houses or where, in the smaller ones, high temperatures are required.

The capacity or performance of a heater is also ascertained by the number of B.T.Us per hour the burners are capable of (B.T.U. stands for British Thermal Unit which is the quantity of heat required to raise the temperature of 1 lb of pure water one degree Fahrenheit). Several of the smaller oil heaters have a B.T.U. output of approximately 2,900 whereas the larger models have a B.T.U. output of 5,800. Certain types are so adaptable that the heat output can be increased simply by replacing the existing burner with a more powerful one.

The amount of paraffin required to keep a heater working will depend on its size and capacity of its burners. Here are some examples: 2,900 B.T.Us, 60 hours burning on one gallon of paraffin; 3,500 B.T.Us, 54 hours and 5,800 B.T.Us, 34 hours per gallon. Some of the small heaters will burn for approximately 120 hours per gallon of fuel. It should be borne in mind, however, that the fuel tank capacity of some of the small types is only 3 or 4 pints.

The correct burning of an oil heater depends on the amount of oxygen it receives. If a greenhouse is completely shut down it is quite possible that the burners will not receive sufficient air and consequently may produce fumes. A very small amount of air can be supplied if a ventilator is opened not

An Aladdin greenhouse heater burns paraffin to warm the atmosphere.

more than ½ inch on the sheltered side of the greenhouse. It is also necessary to keep the wick trimmed regularly and a high-grade oil used.

The positioning of the heater is important if the greatest benefit is to be gained. In a small house the heater can be placed at the far gable end, away from the door and on the central path. It is essential that the heater is placed on a level foundation so that the wick receives a regular amount of oil and also that the fuel gauge indicates accurately.

In a larger house it will be necessary to place the heater in the centre of the greenhouse and on the pathway. To maintain higher temperatures in the larger houses two heaters may have to be used, placing one a little way away from the plain gable end and the second one about halfway along the path. Avoid at all costs a direct draught to a heater as this could cause the flames to flare and set the apparatus on fire.

Oil heaters demand regular attention to filling, wick trimming and adjusting. A heater should be checked about 20 minutes after it has been lit, as it is quite possible that the flames will have increased a little from the first setting and it will be necessary to readjust their height.

Oil heaters are invaluable in the garden shed where they will give frost protection to tubers (e.g. dahlias or potatoes), which have been placed there for winter storage. They also add considerably to your comfort when you are busy in the potting shed during the cold, early part of the season.

It is now possible to provide a measure of automation to oil heaters by feeding fuel from a large drum by gravity to the heater's supply tank via a length of pipe. The large drum can be placed outside the greenhouse and the feed pipe taken inside.

Solid fuel This is a very popular system of heating a greenhouse: it provides the maximum amount of heat for a low consumption of fuel. Like the paraffin heater, however, it is necessary to attend to it regularly. Great advances have been made with designs and many labour-saving gadgets have been introduced.

Many modern boilers have automatic temperature control and simple grate shaking or cleaning levers which facilitate easy management. The layout for a solid fuel boiler installation is quite simple and can be carried out by a novice provided he studies and carefully follows the instruction supplied. The boiler is usually placed outside the greenhouse and often has its own little shed or shelter. There are several models which require no such protection and are either situated just outside the house or built into the side. From the boiler, two rows of pipes (usually 4 inches in diameter) are taken inside the greenhouse and fastened to the sides close to the ground. The pipes terminate in an expansion box.

The system works in this way. The fuel inside the boiler (usually coke) burns fiercely and heats water in a special water jacket which is built into the sides of the boiler. The hot water circulates through the top pipes in the greenhouse and returns to the boiler by the bottom pipes to be heated again. The expansion box, as its name implies, allows for the expansion or contraction of the water as it heats up or cools down.

The amount of piping required and the size of boiler will depend on the volume of the greenhouse and the minimum winter temperatures required. Usually one double row of pipes down one side of a greenhouse is sufficient to provide adequate frost protection. For greater heat and higher temperatures pipes can be extended around the far gable end and down the other side.

Although solid fuel boilers require stoking and cleaning, some of the latest are designed to burn unattended for 14 hours or more. Their fuel consumption, too, is surprisingly economical for the amount of heat generated. For a temperature of about 50°F (10°C) in an 8 foot × 6 foot greenhouse only ½ cwt of fuel is required weekly. Built-in thermostats considerably improve the performance of solid fuel boilers.

Smokeless zone restrictions do not affect the modern greenhouse solid fuel boilers as they can burn most of the recognised smokeless fuels. One problem that arises with these boilers is that of fuel storage. In large gardens it may be possible to allocate an area conveniently close to the boiler for a fuel dump. This enables you to buy in all the fuel you require for the winter period, often at reduced summer prices. The problem of room in the smaller gardens might be troublesome and you may have to have a standing order for fuel so that you receive small amounts regularly.

A Camplex thermostatically controlled glasshouse fan heater.

The maintenance of solid fuel equipment is very easy as there are few parts which are liable to cause trouble or wear out. Everything about these systems is robust and the thermostatic controls are very strongly constructed and extremely simple. Pipes are made from steel or high duty aluminium. The pipe joints are easy to connect by means of special expansion joints which are simply bolted tight when in position. Pipes are usually fastened to the walls or sides of the greenhouse by means of special holders. In this way they occupy the minimum amount of room.

Electricity Electrical heating scores heavily over other systems if complete automation is required. It is, however, one of the most expensive methods to install and run.

Before any form of electric heating can be installed it is necessary to bring the supply of electricity to the greenhouse site. This can be costly if the greenhouse is situated some way away from the source of supply. When siting a new greenhouse it is important to bear this point in mind. There are two ways in which the supply can be brought to the greenhouse; by underground cable or overhead. The former is the best method as the cable is unobtrusive and safely out of harm's way. Underground cable is specially protected against mechanical and chemical damage. It is expensive and should be installed where it cannot be damaged by garden tools. Usually the cable is buried beneath the lawn or close by a path. In certain districts it is permissible to take the mains cable overhead or against a wall, but you must seek the advice of the local electricity authority beforehand.

Where maximum internal working room is important, tubular electrical heaters are ideal as they are fastened to the sides of the greenhouse. Banks of tubes can be quickly installed to maintain any desired temperature. Usually a hand-operated or preferably a rod-type thermostat should be wired to these heaters to ensure automatic and economic running.

Fan heaters can be moved around and are extremely efficient in that they blow warmed air to all parts of the greenhouse. This system ensures that there are no cold corners in the greenhouse. During the summer months the heating elements can be switched off and the fan used for air circulation only. Most fan heaters have a thermostat built in and although not as sensitive as the rod type they ensure reasonable running costs. A fan heater should be placed near the door and at least 2 feet away from the side or other obstructions. If there is the possibility of drips of water from the staging, the heater must be placed away from it. In a small greenhouse this type of heater may take up valuable working room, although it could be placed to one side while work is carried out. It is also necessary to make quite sure that there are no plants in the direct line of the hot air.

Convector heaters draw air in at the base, warm it and send it out hot at the top. They take a little longer to heat up the greenhouse as their warmth is concentrated just round the unit. This heater should be situated at one end of the house if it is a small one, or towards the centre in the larger ones.

A great deal of expense can be saved if small areas of the greenhouse can be heated. This can be accomplished in several ways, one of which is to partition part of the greenhouse and install heating in it. This section can be used

Soil warming cables provide warmth in this mist propagating frame.

as an intensive propagation section. If a small cabinet is made, air heating cables can be installed. These are specially designed and can be connected to a thermostat to provide automatic running.

Electric propagators are available in many sizes and these, too, are an ideal form of confined heating. Usually these propagators are made from easily cleaned fibre glass and the heating is supplied by a length of heating cable which in turn is connected to a control box and thermostat. High temperatures of 65–80°F (18–27°C) can be maintained easily and economically.

References have been made in this chapter to thermostats. The inexpensive types are suitable for loads from 1200 to 2000 watts. They can be regulated to control temperatures between 35 and 75°F (2–24°C). These are hand operated models and are usually accurate to within 4–6°F. The rod types are more expensive but very much more accurate. Usually there is only a differential of 1°F. These models are capable of controlling loads up to 4000 watts. The temperature range is usually wider, 30–90°F. A thermostat must be used with any form of electric heating so that the equipment can be left unattended and running costs kept to a minimum.

With the advent of central heating it is possible for the gardener to make use of the domestic supply if a lean-to greenhouse is purchased. The greenhouse can be placed against a warm sunny wall of the house and a radiator or two can be taken into the greenhouse from the domestic supply. It will be necessary to damp down the greenhouse floor frequently if the floor of the greenhouse is concrete. Central heating is a dry type of heat and a humid atmosphere must be provided to ensure a good growing condition.

Management In a greenhouse you are able to control the 'climate' to a great extent and provide the most desirable conditions for the type of plants being grown. You do this by providing a suitable temperature, ventilation, moisture and shade as and when required.

Without a heating system a greenhouse can be difficult to manage, particularly in the winter. Damp, stuffy conditions must be avoided when temperatures are low and the roof ventilators should be opened whenever possible. Watering should be done very sparingly and it is better to keep the soil slightly dry rather than wet.

A heating system is of value not only in maintaining the desired temperatures but also in preventing excessively damp conditions from developing, particularly when the weather is very damp. For instance, in early autumn, when the air is often very damp at night, a little heat with some roof ventilation will help to overcome the dampness.

In hot weather in the summer the situation is reversed. To prevent temperatures rising too high inside the greenhouse the ventilators should be opened fully and some shade given with blinds or a special distemper applied to the outside of the greenhouse. In addition, water is sprayed on the floors and stagings to create a humid atmosphere around the plants—hot and dry conditions encourage red spider mites and plants lose moisture too rapidly. Watering should be done freely, particularly with plants that have well filled their pots with roots.

Soil composts To grow plants well they need good soil. This is not simply ordinary soil from the garden; it should consist of a mixture of fibrous soil—preferably obtained from decaying turves—moist granulated peat and coarse sand with fertilisers added. Well known John Innes seed and potting composts (see below) have been scientifically devised for the cultivation of plants under glass and, made up strictly according to the formulae, they give very good results. Soil-less composts are being used more extensively these days since they do away with the need for sterilising the loam and their value does not depend upon the quality of the loam as does the value of soil composts.

There were at one time innumerable, highly individual soil composts, especially where potting was concerned, but some years ago the John Innes Horticultural Institution at Bayfordbury in Hertfordshire, decided to seek some standard formulations.

After considerable work and research by Mr W. J. C. Lawrence, a series of standard potting and propagating composts was developed, making it clear that apart from certain modifications in special cases, such standard mixtures would work efficiently for an extremely wide range of plants. And they came as a very welcome simplification of what had been a highly complex, not to say hit and miss process; satisfactory for the skilled gardener, perhaps, but certainly not for those with little experience.

Five composts were devised, three for the various stages of potting, and one each for seed sowing and propagation by cuttings. The primary ingredients are partially sterilised loam—steamed at 200°F (93°C) for twenty minutes, essential in order to destroy pests, diseases and weed seeds—granular peat with a minimum of dust, and coarse river sand up to $\frac{1}{8}$ inch aggregate; with nutrient materials in the form of hoof and horn, $\frac{1}{8}$ inch grist (13 per cent nitrogen), superphosphate of lime (16 per cent phosphoric acid) and sulphate of potash (48 per cent pure potash) and, finally, ground limestone or chalk.

For potting composts Nos. 1, 2 and 3, the basic formula is:

7 parts by loose bulk of medium loam—partially sterilised
3 parts of granulated or moss peat
2 parts of coarse sand.

It is important to note that in each instance the measurements are by *loose bulk* and not by weight. The fertilisers and limestone or chalk are measured by weight, not by bulk.

These materials are mixed most efficiently by placing them in three complete layers, one upon another, and then turning this flat, sandwiched heap over, working from one end to the other and shovelling from ground level throughout. The heap should be turned at least twice—out and back—for maximum success.

The loam may first need to be sieved, where small plants are to be potted, though a $\frac{1}{2}$ inch sieve is fine enough. Where larger plants are concerned, riddling may well be dispensed with, provided the loam is well chopped down.

John Innes potting compost No. 1 is made by adding to every bushel of the combined loam, peat and sand, 4 oz of fertilizer mixed in the ratio of:

2 parts by weight of hoof and horn
2 parts by weight of superphosphate of lime
1 part by weight of sulphate of potash plus $\frac{3}{4}$ oz of ground limestone or chalk, making sure to mix all thoroughly throughout.

This complete mixture is suitable for most basic potting in small to medium pots, for delicate and hothouse plants and those which are not to be kept for any length of time.

Plants other than delicate and hothouse, when requiring a shift beyond the 4 or $4\frac{1}{2}$ inch pot stage, should be potted into John Innes No. 2; made by adding twice the amount of combined fertiliser and chalk to each bushel of loam, peat and sand. John Innes No. 3 comes from adding three times the amount of these materials and is suitable for the more robust and longstanding plants requiring, perhaps, shifts into 8 inch pots and beyond.

Compost made up for plants objecting to lime should obviously have the chalk or limestone omitted; it should be reduced where lime is present to any extent in the loam used in the basic formula. Ideally, this should not be so, with loam with a *p*H of 6·5, the kind to use if possible.

The John Innes standard seed sowing mixture consists of:

2 parts by bulk of loam—partially sterilised
1 part by bulk of peat
1 part by bulk of sand

to each bushel of which is added only $1\frac{1}{2}$ oz superphosphate of lime and $\frac{3}{8}$ oz ground limestone or chalk. The loam and peat may have to be rubbed through a $\frac{1}{2}$ inch sieve to make them sufficiently fine for seedling growth.

Where many tiny alpines are concerned, extra grit in the compost is often required in order to allow them very adequate drainage, essential to many of them, especially in their early stages.

Cutting composts The compost into which cuttings are put to root must provide a suitable medium for encouraging the quick formation of roots. It is not designed to feed the plants and thus it is good gardening practice to pot up rooted cuttings into a potting compost once good roots are formed.

The essential qualities of a cutting compost are hygiene (freedom from soil-borne pests and diseases), air and good drainage combined with the ability to hold sufficient moisture. Therefore a gritty and porous texture is best.

Silver sand is the most reliable rooting medium for most cuttings but does not provide any plant food at all and so the cuttings once rooted must be moved on to a compost containing plant nourishment. Sand and peat mixed in equal parts probably gives the best results and gives the gardener a better margin of time before potting up is required because this mixture does provide food.

The John Innes cutting compost is composed of:

1 part by loose bulk of medium loam
2 parts of peat
1 part of coarse sand.

This compost can generally be bought from sundriesmen ready made up. Proprietary rooting media such as vermiculite, No-soil compost and Rootine are also available from sundriesmen and all ensure good results.

Soilless peat-based composts These new seed and potting composts are already widely used commercially and may well replace the older John Innes mixtures. Experimental work in recent years shows that tomatoes, cucumbers, many garden flowers and even cacti grow well in peat-based composts containing plant nutrients. These are proprietary composts developed by leading horticultural suppliers. The composts should be used according to the instructions supplied with them. Failures may occur if the gardener omits an initial thorough watering, compresses the compost excessively, or permits the compost to dry out when in use. It is advisable to purchase a fresh supply of soilless composts for each season's work. The

Composts based on peat are superseding those based on loam. 1 Fison's Levington soilless compost is filled into pots without firming. 2 Antirrhinum seedlings pricked out in John Innes potting compost where they will remain until they are ready to be planted out.

An electric steam sterilising unit. The bottom of the bucket consists of a metal mesh. This allows steam to circulate freely through the soil to sterilise it.

To sterilise seed, or potting, compost once it is in the seed tray, water with dilute formaldehyde (1 quart to 12 gallons of water) at the rate of 2 quarts to a square foot.

composts are clean to handle and are recommended for the raising of plants in the greenhouse or in containers in cold frames.

Hygiene Pests and diseases spread very rapidly in the warmth of a greenhouse and a careful watch should be kept for the first signs of their presence. Fortunately many can be controlled effectively by using insecticidal or fungicidal smokes and sprays.

Many greenhouse gardeners prefer to use the same soil year after year for tomatoes, cucumbers and other crops grown in borders. Soils used continuously in this way are almost certain to become infected with eelworm and organisms that cause root diseases. These must be destroyed if good growth is to be ensured, or else the soil has to be changed about every three years—a very laborious and often expensive job.

Loam used for seed sowing and potting must be free from all harmful organisms and substances and sterilisation will help to do this.

How to sterilise your soil You can either heat the soil or treat it with chemical sterilants. The latter are more widely used by gardeners because they are more convenient and do not require expensive equipment.

Most sterilants available to gardeners are liquids, which when applied to the soil change to gases so that the killing action is extended for some distance beyond the point of application. Soil sterilants are sold under a variety of trade names. They may be any one of four or five chemicals in common use for this purpose.

Chemical sterilisation is not so thorough as steam heat treatment which when properly applied eliminates all pests, diseases and weed seeds. Chemical sterilants are selective in their action. For example, formaldehyde is more effective against diseases than pests, and cresylic acid kills pests rather better than it does fungi.

How to get the best results from chemical sterilants:

Before use *Moisture content* Soil with proper moisture content should just hold its shape when moulded in the hand and crumble readily when dropped. If too wet the sterilant fails to diffuse through the soil.

Temperature of soil The best range is 60–80°F (16–26°C). Below 60°F (16°C) sterilants kill more slowly, while at temperatures above 85°F (29°C) most preparations leave the soil too rapidly to be completely effective.

Conditions Cultivate thoroughly to a depth of 6 to 8 inches, adding peat or crumbly manure or compost prior to sterilisation. The prepared soil should be free from clods, lumps or unrotted manures.

Rotary cultivators leave the soil surface fluffy and allow the sterilant to escape too quickly. If a tiller is used allow soil to settle for a few days before application.

Soil type Best results are obtained in sandy soils. Soils high in clay and humus may require a higher dosage rate.

During use The job must be done thoroughly if you are to get full benefit from your sterilant. All the soil must be

Cover immediately with plastic sheeting and secure it well.

properly saturated or mixed with the chemical, otherwise pockets of infection will appear in patches neglected during application.

Particular care should be taken to apply the chemical thoroughly round the walls and supporting posts since pests and diseases harbour in cracks and crevices. A clean brick face is more easily washed down than one in which the mortar is cracked; crumbling and cracked walls should be repointed before washing down.

The inside structure of the greenhouse provides good harbourage for pests and diseases and enables them to live from one season to another; so this should be sprayed or washed down with a sterilant before you start on the soil.

After use *Aeration* Allow sufficient time for the fumes to escape from the soil before you plant. Check for presence of sterilant by holding a handful of soil, from a depth of 6 inches, near the nose and sniff. A further useful test for the presence of residues is to take a few handfuls of soil from several different places a few inches deep. Mix thoroughly, moisten if necessary and half fill a jam jar with the mixture. Sprinkle cress or radish seed on the soil and put in a warm place with the top on. If the seeds germinate normally it is safe to plant; if they do not, the soil should be stirred again. Clearance can be speeded up by forking or cultivating the soil about 3 weeks after application, or, in a greenhouse, by putting on the heat.

Recontamination Sterilised soil can be reinfested by untreated soil being brought into the greenhouse on tools or shoes. They should be dipped in a dilute solution of the sterilant before entering the greenhouse.

Seed and potting composts When treating loam for seed and potting composts add to one bushel of soil (which equals about 1 cubic foot) $1\frac{1}{4}$ gallons of dilute solution of formaldehyde ($\frac{1}{4}$ pint formaldehyde to $1\frac{1}{4}$ gallons of water) and mix thoroughly with a spade. Cover the heap with a plastic sheet for 48 hours. Uncover the heap and spread soil out to dry. Turn the soil several times during the next few weeks to clear fumes. Do not plant for at least 4 weeks. This operation should not be done anywhere near growing plants.

Seed trays Use prepared seed or potting compost, fill the seed trays and apply about 2 quarts of diluted formaldehyde

---

Some of the many varieties of Fuchsia in a greenhouse at Wisley.

per square foot of seed tray. Stack seed trays and cover with plastic sheet. Remove sheet after 3 days, and air thoroughly until all odour has disappeared before planting.

Sterilisation by heat For a few diseases heat treatment is still the only method of control. The commonest and in many ways the best method of sterilising soil by heat is steaming. Steam-generating plants are widely used but are not primarily intended for small greenhouses. You can buy steam sterilisers that consist of galvanised water tanks fitted with metal soil containers, which fit tightly over a water trough electrically heated from below.

Trays of seed potting compost can be treated by drenching thoroughly with boiling water. Very small quantities of soil can be steamed by first boiling ½ pint of water in a saucepan, then adding several pints of air-dry soil, covering tightly and boiling for 10 minutes. The soil should be allowed to cool for a further 10 minutes before removing the cover.

For the quickest results with the least steam, the soil to be sterilised should be nearly dry. If moist much more steam is required to raise its temperature to the sterilisation level.

As an alternative to steam, electrode sterilisers are used. These consist of a metal box containing an electrical heater usually of the flat plate type, and the soil is placed in the box and packed around the plates. As the electric current passes through the soil its temperature rises. For this type of heat treatment, the soil should be moist and a temperature of only 150°F (66°C) is sufficient to kill practically all the organisms in it.

Electricity in the greenhouse

Apart from heating the greenhouse, as described in chapter 3, electricity can be used in various ways in the greenhouse.

Electricity is the only system which provides complete, trustworthy automation for several applications. It is clean, relatively easy to install and the installations are neat in appearance. Electrical apparatus is safe only if it is installed by an experienced electrician and only if the correct or approved types of fittings are used. There is a high level of dampness in the garden and in most horticultural structures and equipment and it is vital, therefore, that waterproof fittings and cable are used. Local Electricity Boards are always pleased to offer sound, practical advice on proposed installations and will also undertake the work.

Soil warming This is provided by specially manufactured heating wires or cables which can be either low voltage

A modern greenhouse in which electricity is used for heating, ventilation, soil warming, lighting and shading. The aim is for complete automation.

or mains voltage. The former uses a transformer which reduces the mains voltage to a low, safe level of about 12 volts. The heating element is usually galvanised wire, covered with plastic, of a given length and gauge which will provide a calculated electrical resistance. Elements of different lengths can be purchased so that any given area can be warmed to a certain temperature. Usually a loading of 6 watts per square foot is sufficient.

Mains voltage warming uses a special wire covered by insulation and, in the more expensive types, a protective sheath. Unlike the low voltage outfits, the wires are connected direct to the mains power supply.

The approximate running costs of soil warming are ½p per week for each square foot of bed which is warmed. For a frame of 32 square feet the cost per week would be about 14p if a dosage of 10–12 hours each night was allowed. Costs will rise, of course, if higher temperatures are required.

Hot beds are formed quite simply. The heating wire must be laid over the base of the bed in even patterns, usually in parallel rows about 4–6 inches apart. The wires can be retained in position by special wire pegs. The required loading for a given area is ascertained by multiplying the area by 6. For example, a hot bed 8×4 feet has an area of 32 square feet. Multiplied by 6 the loading required is 192 watts. A 200 watt cable or transformer would be correct.

Heating for a hot bed is usually given by the dosage method during the night. A certain number of hours of warming is applied, usually 12 hours in the colder northern gardens and 10 hours in the more favoured southern areas. Automatic control can be used in the form of a special switch which will switch the electricity on and off at pre-set times.

Once in position, the warming wires are covered with soil to a depth of 6 inches. For complete safety, the hot-bed system is recommended as little damage will occur if the wire is accidentally cut by a spade, fork or trowel. The cost of units will vary according to the type. A non-automatic type for an area of 32 square feet costs approximately £15.25. An automatic system is surprisingly cheap and would cost about £9 for a 20 square foot area.

Propagation In the greenhouse or frame propagation is made much easier by the use of electricity and a greater degree of temperature control is possible so that growing conditions are as near perfect

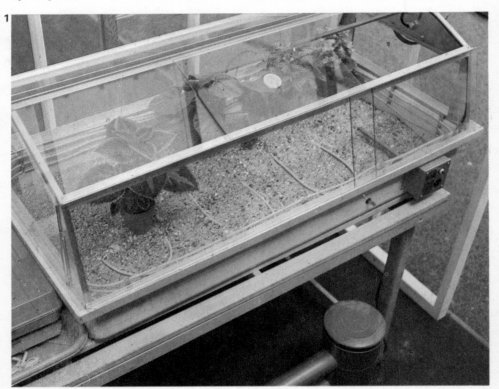

1 The Humex Big Top Propagator has thermostatically controlled bottom heat and air warming.
2 A close-up of the soil/air thermostat used in connection with mains warming cables or transformer units. This ensures automatic control of temperature.

as possible. A home-made propagator can be constructed from deep boards about 1 inch thick and 6–8 inches in depth. The top of the 'box' is covered by sheets of glass or clear plastic sheeting. The box need not have a base if it is placed on the greenhouse bench.

A higher soil temperature is required for this type of work and temperatures of 55–60°F (13–16°C) are necessary. This means that the loading per square foot of propagator area must be increased, but only slightly, by 1½ watts to 7½ watts per square foot. A propagating case of 6 × 2 feet is more than sufficient for average purposes. The cables should be laid over 2 inches of washed sand and covered with 3 inches of sand. The pots, boxes or seed pans are placed on the sand and warmth is retained by packing granulated peat around the receptacle afterwards.

A further refinement is the use of cables which will warm the air inside the propagator. These are clipped neatly in place around the sides of the frame by means of special stainless steel spring clips. These cables need not produce a very high temperature when used in conjunction with soil or sand warming and a temperature of 50°F (10°C) is adequate for normal purposes. Automatic control can be achieved if rod-type thermostats are used, one in conjunction with the cables in the sand, and one above for air temperature control.

Commercially-made propagators of various designs and sizes can be obtained. The size is determined by the amount of use there is for them. Some models are supplied with a special plastic dome as an extra to give more headroom. Most modern propagators have plastic or glass fibre bodies which ensure trouble-free service. Many are equipped with accurate thermostats which regulate the temperature of the sand bed and ensure low running costs. Heating is supplied by warming cables which are laid on a sand bed and covered by another layer of the same material. Seeds are sown in pots, pans or boxes which are placed on the warmed sand and surrounded by granulated peat to retain the warmth. Cuttings can be treated in the same way. The top of the propagator is covered by glass or plastic. In this way soil temperatures of 60–65°F (16–18°C) can be maintained and concentrated in the small area of a propagator instead of heating the vaster area of the greenhouse. This is a great advantage when it is realised that the running costs of a propagator are 2½–4p per week.

Very simple and much smaller propagators can also be purchased which are really 'heating plates'. These provide bottom heat and on them the boxes or pots are stood. Some of the latest models have their heating elements and thermostats bonded in completely by a special plastic process. This ensures a very neat, compact unit which is completely safe and rugged. The latest propagator is supplied with an aluminium glazed top. It has air warming and bottom heating cables both independently controlled by accurate thermostats. The running costs when both are in use amount to only 6p per week.

Pest control Another method of dealing with greenhouse pests by electricity is with a special 'lamp' which volatilises insecticides. Tablets are inserted through a hole in the lamp and the heating element volatilises them.

Automatic ventilation One of the most serious problems is that of keeping the greenhouse cool. Electrically operated fan extractors are useful here, particularly for the gardener who is away for most of the day. These are attached to the gable end of the greenhouse opposite

1 Capillary watering of plants is automatic. This system can be used safely in conjunction with soil-warming cables.
2 A thermostatically controlled fan can change the air 30 times per hour.

to the door and as high up as possible. Connected to a thermostat, the extractor fan comes into operation as soon as the temperature exceeds that which has been set on the thermostat. It is essential to select a fan of sufficient power and diameter to suit the cubic capacity of the greenhouse. For 500 cubic feet a 9 inch diameter fan is usually adequate. For a capacity of up to 1000 cubic feet a 12 inch diameter fan will be necessary.

There is another ventilation system which consists of a powerful, thermostatically-operated electric motor, nylon cords and pulleys. These are so arranged that they open and shut the ventilators by their handles. The motor unit winds or unwinds a spindle, attached to which is the nylon cord. This system takes a little time to assemble and test.

Automatic watering Electricity can even play its part where watering is concerned. A completely automatic system can be installed, which controls the amount and frequency of the watering. The 'brain' of the outfit is a control unit which contains a solenoid valve which switches the mains water supply on and off. When used as a watering device, a detector probe is inserted into a special compartment in the unit which can be flooded with water through a porcelain block. As soon as the porcelain block absorbs all the water from its own basin and the probe's compartment, an electrical circuit is made and the solenoid valve turns on the water. As soon as the block and the compartment are flooded, the solenoid is switched off and the water ceases to flow. The duration of the water application can be controlled by the adjustment of a special screw which regulates the flow of water to the block and probe compartment. Usually the water to the plants is delivered through a perforated plastic pipe connected to the one end of the control unit.

Where the apparatus is required for water application for propagation purposes, it is essential to provide a very small amount and this is done by the use of special mist spray heads. These are usually fixed at intervals to a length of pipe, the heads themselves being about 2 feet above the staging. The detector probe is removed from its special compartment in the control unit and placed on a stick just above the cuttings where it will be wetted by the mist heads. The detector, in such a position, will dry out quite quickly and will provide more frequent watering. The detector is often referred to as a 'leaf' when used in this

Humex tubular heaters are ideal for greenhouses of almost any size or shape. They are easily fixed in position.
The electrical loading is 60 watts per foot of single tube.

way, as it will dry out in the same way as a leaf does.

Lighting Lastly there is the use of electricity for illuminating the greenhouse, not only to enable you to attend to your plants in the long winter evenings, but to encourage out of season growth for certain plants.

Good gardening depends not only on skill but also on patience, but sometimes there is an opportunity to hurry things along. The growth of some plants and their flowering and fruiting times can be influenced by adjusting or varying the number of hours of light and darkness per day. All that is required is low intensity light. Plants can be divided into two types—long-day plants and short-day ones. Increasing the length of day by lighting will induce the flowering of long-day types during short daylight hours. Short-day plants such as mid season and early-flowering chrysanthemums can have their flowering times delayed if a day length of over 14½ hours is provided from mid August to about mid October.

There are available specially designed plant irradiators of various wattages. For the treatment of chrysanthemums, 100 watt lamps are suitably suspended about 6 feet above the rows and spaced 5–6 feet apart. More intense lighting of 400 watts is suitable for raising seedlings such as tomatoes early in the season. Such mercury lamps have an average lamp life of 4000 hours.

For average amateur requirements 100 watt lamps or multiples of these are quite satisfactory. For bulb forcing, one such lamp suspended over daffodils and tulips supplied with an air temperature of 60°F (16°C) will encourage much more rapid growth. Twelve hours dosage per 24 hours will be sufficient and this can be more conveniently regulated if a time switch is employed. About 9 square feet of area can be dealt with efficiently by one of these bulbs which should be placed 3 feet above the bowls or boxes.

Artificial light is ideal for the cultivation of saint paulias and specially designed growing-cabinets can be purchased for this purpose. A special strip light fixed in the roof of an electric propagator will encourage more rapid and vigorous growth of seedlings, especially during the duller early spring months. In all instances, it is *vital* that only specially manufactured apparatus and fittings are installed and the installation should always be carried out by a competent electrician.

Installation kit The exception to the rule of not attempting to install electrical equipment oneself is provided by the 'do-it-yourself' kit which enables you to fit out your greenhouse with a comprehensive array of socket outlets, mains switches and lights. The manufacturer supplies dummy (stiff card) matrices for plug points. These are pinned in the desired position in the greenhouse. The distance between each is then carefully measured so that the necessary amount of cable can be supplied. The distance between greenhouse and dwelling house is also measured, either for underground mains cable or above-ground installation. All these details are entered on to a special order form supplied. Eventually the kit arrives completely wired professionally, including the mineral-insulated, protected mains cable, ready connected to suitable connectors.

So complete is the kit that even the special cable clips are supplied so that an extremely neat installation can be fitted. A kit comprising a lighting unit, six socket outlets, a mains unit, switch fuse and the cable costs about £13.

What to grow in your greenhouse

A very wide range of plants, flowers, fruits and vegetables, can be grown in the greenhouse, depending on the minimum temperature that can be maintained in the winter.
Unheated greenhouses A great deal of interest can be had from a greenhouse with no heating equipment and if you are working to a small budget an arrangement of this sort is a good beginning; heating equipment can be installed later on as you become more experienced and wish to experiment with a wider range of plants.

An attractive display can be had in an unheated greenhouse by growing mainly hardy plants in pots. These will flower a little earlier than those in the open but the flowers will not be spoilt by inclement weather. Other plants that are usually started into growth early in warmth for summer flowering, can also be grown by starting them later when outside temperatures are higher.
Shrubs The camellia is often thought to be tender, possibly because the flowers, which appear early in the year, are sometimes damaged by frost. For this

A well-grown 'Hortensia' Hydrangea.

reason the protection of a greenhouse is valuable. All the varieties of *Camellia japonica* will grow happily in pots of lime-free soil, but they should not be cossetted as if they were hot house plants. *Prunus triloba flore pleno* has double pink flowers and is another fine flowering shrub for the cold greenhouse. The yellow-flowered forsythia and rosy-purple *Rhododendron praecox,* lilacs as well as winter-flowering heathers *(Erica carnea)* will all give a bright display. For flowering in the summer *Hydrangea paniculata grandiflora* produces enormous white flower trusses. It is easily grown in pots and the stems should be cut back severely each spring. The 'hortensia' hydrangeas will flower earlier than the species mentioned above and really good blue flowers can be had if the soil is treated with a blueing powder.

One of the many forms of Camellia japonica.

Bulbs These can provide a wonderful display early in the year. The earliest to flower are 'prepared' daffodils in January followed by 'Paper White' narcissi, the dark blue *Iris reticulata* and named varieties of snowdrop. Other small bulbs well worth growing are miniature daffodils, such as *Narcissus cyclamineus, N. bulbocodium* and *N. triandrus albus,* winter-flowering crocuses and *Eranthis hyemalis,* the winter aconite with buttercup-yellow flowers.

For flowering in the summer, lilies such as *L. auratum* and *L. speciosum* look superb grown in pots. Gloxinias and tuberous-rooted begonias are popular plants and the tubers of these can be started into growth in early April.

Other flowers for a spring display, which can be purchased or lifted from the open garden in the autumn and potted up for the cold greenhouse, are wallflowers, dicentras, astilbes, forget-me-nots, polyanthus, lily-of-the-valley and Christmas roses *(Helleborus niger).*
Fruit and vegetables Even in a cold greenhouse, grapes will succeed but it is important to choose suitable varieties such as 'Black Hamburgh' or 'Foster's Seedling'. The vines can be grown in

large pots or tubs with the stems trained up a supporting framework, or they can be planted permanently in a border of good soil.

To have tomatoes fruiting under glass in June the seed has to be sown soon after Christmas in a hot house. They can also be grown in an unheated greenhouse but fruit will not be ripe until late summer. Young plants can be purchased for setting out in a border at the end of April.

There is no reason why melons and cucumbers should not be grown in a greenhouse without artificial heat. As with tomatoes, early fruits cannot be expected but all the same they taste much better when picked fresh from one's own garden.

Peach trees can be grown in the open garden but they flower early in the year and their blossoms are likely to be damaged by spring frosts. Given the protection of a greenhouse this hazard can be avoided. Peaches make luxuriant growth and they can easily 'swamp' a greenhouse. For this reason they should be grown as fan-trained trees with all the stems neatly trained out up the roof or against one end of the greenhouse.

1 Melons are among the fruits that may be grown in a greenhouse even without artificial heat.
2 Tomatoes ripening in an unheated greenhouse.

This small greenhouse is filled with colourful pot plants. Full use has been made of the available space by suspending pots from the roof.

Good varieties for a cold greenhouse are 'Peregrine' and 'Duke of York'.

Where there is a vacant border of soil in an unheated greenhouse during the winter, lettuce can be grown. Seed may be sown in October, and will produce seedlings ready for planting in November or December, and they should be fit for cutting in April and May.

It must not be thought that all these flowers, fruits and vegetables can be grown together in a small greenhouse. If you prefer flowers it is best to grow little else although a vine or a peach tree could be grown with them. Tomatoes do not like heavy shade and so it would be unwise to have a vine, which produces a heavy coverage of foliage in the summer, with them. Cucumbers and tomatoes should also be kept separate for the best results. Tomatoes like plenty of light and air in hot weather, whereas cucumbers like tropical conditions—heavy shade, high temperatures and high humidity.

Cool greenhouses Although great interest can be had from an unheated greenhouse a wider range of plants can be grown and earlier fruit and vegetables can be had if the greenhouse has a minimum winter temperature of 50°F (10°C). With the aid of artificial heat it is also easier to maintain a good growing atmosphere or climate for the plants.
Shrubs As in the cold greenhouse, hardy shrubs such as forsythia and *Prunus triloba* can be brought into the greenhouse in winter for flowering much earlier than those in the open. Camellias will thrive, provided the temperature is not allowed to shoot up too high in the day—this causes the flower buds to drop—and hydrangeas in pots can be made to flower early. Indian azaleas are popular florists' pot plants. These can be kept from year to year in a heated greenhouse, provided they are fed regularly. They flower in winter and early spring and can be put outside for the summer. *Acacia dealbata,* or mimosa, with yellow fluffy flowers and a heady scent can be enjoyed in early spring. There are also a great many ornamental climbers that will enjoy the warmth. The passion flowers, the brightly coloured bougainvilleas, the soft blue *Plumbago capensis* and *Lapageria rosea* with rose pink, waxy bells, all flower in summer.
Bulbs All the popular bulbs such as daffodils, hyacinths and tulips, can be made to flower in the dark days of winter in a heated greenhouse, but after the pots and bowls are removed from the plunge beds they should be given cool conditions at first and gradually acclimatised to warmer conditions.

Hippeastrums, often mistakenly called amaryllis, have large, handsome flowers and the bulbs may be started into growth in February for spring flowering. Freesia corms, started into growth in

**1 Gloxinias can easily be raised from seed for flowering in the cool greenhouse.
2 Begonias too may be raised from seed.
3 Regal Pelargoniums are excellent plants for the cool greenhouse.**

August, produce their colourful and scented flowers in February and March; they will grow well in a minimum temperature of 40°F (4°C).

Arum lilies are not true bulbs—they have tuberous roots—but in a heated greenhouse they will flower in the spring. An easily grown and handsome bulb is the Scarborough Lily, *Vallota speciosa*. It can be stood outside for the summer and in August it will produce its vermilion trumpet flowers on stout stems. Flowering a little later, nerines have delightful, glistening flowers in pink, red and white; they differ from many bulbs in that they need to be rested and kept dry in the summer.

Begonia and gloxinia tubers may be started into growth in March for flowering in the summer. Achimenes can be treated similarly, grown in pots or in hanging baskets.

Other flowers that will flourish in a greenhouse with a temperature of 50°F (10°C) in winter are perpetual-flowering carnations—they like light and airy conditions and will flower for most of the year; chrysanthemums for autumn and winter flowering; and fuchsias for the summer.

Pot plants that can be raised from seed in spring and early summer for a display in winter and spring are: *Primula obconica, P. malacoides* and calceolarias. Some gardeners, unfortunately, are allergic to *P. obconica* and if they handle plants it sets up an unpleasant skin irritation.

Cyclamen can be grown successfully from seed sown in August, to provide plants for flowering 16 months later. The poor man's orchid, schizanthus, is also easily raised from seed in August for flowering the following spring.

Regal pelargoniums are becoming more popular and they are useful for their handsome flowers borne from June onwards; these are best propagated from cuttings taken in late summer.

Fruit and vegetables Cucumbers and melons need a temperature no lower than 60°F (16°C) to grow well and it is not wise to make sowings before April. Tomatoes also need a temperature of 60°F (16°C) for germination and seed should not be sown before early April in the cool greenhouse.

Peaches and nectarines can be started into growth early with a little artificial heat but it is important to remember that the trees are given a rest in cool conditions in the winter. Suitable varieties are: Peaches—'Early Rivers', 'Royal George' and 'Hales's Early'; Nectarines—'River's Orange' and 'Early Rivers'.

In a moderately heated greenhouse grape vines can be started into growth early and a little heat in early autumn helps to complete the ripening of the grapes. Suitable varieties are 'Black Hamburgh', 'Foster's Seedling' and 'Madresfield Court'.

Given good light, lettuce is a useful winter crop in a moderately heated greenhouse. From a sowing in mid-October the seedlings are pricked out in boxes and planted in prepared beds in November; the lettuce should be ready for cutting in February. A suitable variety is 'Cheshunt Early Giant'.

In addition to those already mentioned, other vegetables and salad crops well worth growing in a cool greenhouse are

1 Grape 'Black Hamburgh' is a popular variety for the greenhouse.
2 A well-stocked greenhouse in the Botanic Garden, Glasgow.

mustard and cress, radishes and aubergines. Rhubarb and seakale can also be forced under the greenhouse staging in complete darkness.

Warm greenhouses To be able to heat a greenhouse to a minimum temperature of 55°F (13°C) on the coldest night in winter is an expensive undertaking but it is also rewarding. A great many fascinating tropical or 'stove' plants can be cultivated in addition to those already mentioned.

Shrubs Among the shrubby plants with handsome foliage for a warm greenhouse, most of which are rarely seen, codiaeums (crotons) are outstanding. They have leaves of various shades, brightly marked with green, red, yellow and orange. Dracaenas also have attractively coloured leaves and they include *D. godseffiana* with green, white-spotted leaves and *D. fragrans victoriae* with long green and yellow striped leaves.

The poinsettia, *Euphorbia pulcherrima*, a popular plant at Christmas time with red rosettes of bracts, needs a warm greenhouse to grow well and so does its close relative *E. fulgens*, which has small orange-red flowers on arching stems in winter.

Gardenias, which are prized for their pure white fragrant flowers, do best in a warm greenhouse. Less commonly seen is *Brunfelsia calycina*, an evergreen shrub with purple flowers.

For training up the greenhouse roof there are numerous exotic climbers. The Madagascar jasmine, *Stephanotis floribunda*, has thick leathery leaves and clusters of white scented flowers. There are also several clerodendrums with colourful flowers; *C. thompsoniae* has crimson and white flowers and *C. splendens* produces clusters of red flowers.

Bulbs Begonias, gloxinias, hippeastrums and smithianthas can all be started into growth in January or early February; seed of begonias and gloxinias can also be sown in January. Apart from these popular types, *Eucharis grandiflora* with beautiful, white and fragrant flowers will revel in a warm greenhouse. The tuberose, *Polianthes tuberosa*, is

another bulbous plant well worth cultivating for its white, fragrant flowers. Caladiums have tuberous roots and they are grown for their handsome foliage. The tubers can be started into growth after resting in the winter.

Other plants that enjoy a warm greenhouse include coleus and *Begonia rex*, both of which have highly ornamental foliage. Winter-flowering begonias, provide a wonderful display of colour in white, pink and red. Saintpaulias, so popular as room plants, do best in a well-heated greenhouse.

Fruit and vegetables Most of the kinds already mentioned can be grown in a warm greenhouse. Cucumbers and melons enjoy high temperatures and a humid atmosphere. Peaches and nectarines do not require high temperatures and the same applies to vines, although

1 'Bellegarde' is one of a number of varieties of Peach which do well in a greenhouse with slight heat.
2 'Royal Sovereign' is a good Strawberry variety for forcing under glass to obtain early fruit.

that very fine grape, 'Muscat of Alexandra', a tricky variety to grow, enjoys warmer conditions, particularly at flowering time and when the berries are ripening. Strawberries, put in pots in late summer will produce early fruit if they are gradually introduced into warmer conditions in February. 'Royal Sovereign' is a good variety for forcing. Figs are not often grown under glass but they can be cultivated in pots and, gradually given warmth in January and February, will produce ripe figs in late spring.

Pot plants

A very large number of flowering plants are raised each year in pots to be sold for the decoration of rooms, shops, etc. The quantity produced and the selection of genera and species to be grown are governed more by commercial than horticultural considerations. This means that many of the most delightful plants are not grown because they are either unsuitable for room decoration or are too expensive to produce. However, in keeping with the law of supply and demand, more of these plants are grown today.

There is now, for example, a very extensive Christmas trade in poinsettias; plants which not so many years ago were the prerogative of those gardeners who had the advantage of a heated greenhouse. The poinsettia is an example of the fluctuations in popularity due to the qualities that growers require in a pot plant that is destined for room decoration. As originally introduced the poinsettia was a rather tall plant, which showed great susceptibility to draughts. The Edwardians would put them in their jardinieres, but a few days after being brought into the house all the leaves would fall off leaving a straight, bare stem except for the rosette of scarlet bracts at the top.

As the plants have such brilliant flowers at the darkest time of the winter efforts were made to obtain shorter plants which could be used for table decoration. For this purpose cuttings were taken rather late in the season and the plants put on shelves in the greenhouse, suspended near the glass. In the days when people of even moderate means could afford gardeners, a man could be running up and down ladders half the day to water the plants that were often placed 10 feet from the ground. Nurserymen would also grow these rather expensively produced plants, and they cultivated comparatively dwarf plants. The plants still showed susceptibility to draughts and they would often be put into rooms that might be roasting hot at night, but barely frost-proof during the day-time. Heating generally came from an open, coal fire and this would often only be lit in the evening or

**Hippeastrums are bulbous plants, often erroneously called Amaryllis.
Their bright flowers make them attractive as room plants.**

afternoon. Under these conditions the poinsettia was not really a very popular commercial plant.

As the twentieth century advanced the use of central heating became more widespread, particularly in North America, where the winters are so cold that open fire heating was quite inadequate. The atmosphere produced by central heating is more congenial to tropical plants, such as the poinsettia. On the other hand labour and heating both became more expensive and it was no longer commercially practical to produce dwarf plants in the old way. The poinsettia was sold as a rather tall, lanky plant.

At the same time a certain amount of selection was going on to find a plant that would be less susceptible to draughts and to violent fluctuations of temperature. Even if the room were draught-proof, the plant had been brought from the nursery to the shop and from the shop to the purchaser's home, and this would defoliate the poinsettia through a succession of chills and heat. There was some success in this selection to find a plant which would resist draughts, and before the Second World War there were strains of poinsettia which were fairly retentive of their leaves.

The next step came after the war with the discovery of chemicals that would dwarf a plant by simple application. It was this that really transformed the situation and enhanced the commercial

The 'Mikkelsen' strains of the Poinsettia, Euphorbia pulcherrima, make flamboyant and colourful pot plants.

1 Kalanchoë 'Vesuvius' is a succulent bearing red flowers in winter.
2 The Calceolaria hybrids, sold as C. multiflora nana, provide a wide range of colour and make compact flowering pot plants.
3 Coleus blumei, a commonly cultivated plant, has variegated leaves in bright colour combinations.
4 Saintpaulias, or African Violets, make interesting pot plants and there is a good choice of named cultivars, with flowers in a wide colour range.

possibilities of the poinsettia. It became possible to produce a compact scarlet-flowered plant for Christmas blooming. The selection of the 'Mikkelsen' strain, which will retain its leaves under the most adverse conditions, has resulted in a plant that has almost all the qualities that the grower and purchaser of pot plants requires. It is a plant that is colourful at a time when there is little colour outside, sufficiently compact to be placed on tables without appearing too intrusive, and will maintain itself in good condition without too much attention.

In describing the qualities that the breeders of the poinsettia tried to obtain the desirable features of the pot plant have been shown. It is chiefly in the months between October and April that pot plants are most in demand. They should be fairly easy to manage, which means that they do not require great heat or a very wet atmosphere. They should be compact, because shrubs some 5 feet in height even though they are covered in flowers for two months on end would be far too large for most rooms. The plants should give a good display of blossom, because a plant that has a long-flowering season, but never bears more than a few flowers at a time, will not be very popular commercially. That is one of the reasons why the gardenia has never become very popular as a pot plant; it has a long flowering season, but seldom a very large display of blossom. This might be overcome, in view of the plant's fragrance, if it were slightly more easy to manage.

People will grow difficult plants such as cyclamen and saintpaulias provided the display is sufficiently rewarding, and they will grow easy plants that are less colourful, but they will not take trouble over plants that are neither easy nor very brilliant. Possibly, if there were a demand for a slow-growing, rather difficult plant, such as *Medinilla magnifica*, growers would develop a strain that was comparatively easy to propagate rapidly.

Plants indoors Keeping a plant in good condition in a living-room is more difficult than doing so in a greenhouse. It is usually easy enough to provide suitable temperatures and there are plants to suit most conditions. The difficulties that are to be faced are more concerned with light and with atmospheric humidity. Plants in nature or in most greenhouses (lean-to's) are an exception, and receive light from all sides and from above. Plants in rooms are generally illuminated on one side only.

Plants in the wild, and in a well-managed greenhouse, grow in a humid atmosphere. The amount of humidity depends on either rainfall or on the way the greenhouse is damped down, but the atmosphere is moist. In rooms, particularly during the winter months, the air tends to be dry. Plants that grow in deserts appreciate dry conditions, but hot deserts generally have brilliant light; a combination that is not too easy to reproduce in the greenhouse, and well-nigh impossible in domestic rooms. The choice of room plants is therefore restricted to those that normally like rather shaded conditions and those that survive reasonably well in a rather dry atmosphere.

The problem of providing atmospheric humidity can be partly overcome by plunging the pot, or pots, into some other container, and filling the interstices with any absorbent material, such as peat, moss, vermiculite, sand, or even wet newspaper. This material is kept always moist, so that watered material will constantly surround the plant and contribute the necessary moisture to the air. This is fairly effective, but it is not the same as a uniformly humid atmosphere. In some rooms of the house, such as kitchens and bathrooms, the atmosphere is naturally rather steamy and many plants will thrive better in such situations. Nowadays, when the kitchen often serves as a dining room as well, there is much to be said for having plants in it, but there is usually a space problem in such rooms and this may prevent having your plants there. Although plunging the pots is only partially effective, it is sufficient to enable you to preserve a large number of flowering pot plants.

The problem of providing light can be partially overcome by turning the plant slightly each day, so that every part of the plant is illuminated in turn. Plants such as the Christmas and Easter cactus will resent this and drop their buds, but most flowering plants will appreciate all parts of the plant receiving light. Rooms with a bay window can have their plants placed so that they are illuminated from three sides, but the bay is sometimes rather cold and draughty.

Cool conditions will do no harm to flowering plants, provided that the cold is not excessive, but draughts are often killing, causing the buds to drop before opening and often resulting in defoliation. Plants placed on window sills usually receive sufficient light. However, unless the windows are double glazed the plants should be removed to the centre of the room when conditions are frosty.

Most of the winter-flowering pot plants that are sold are plants that are suitable for the house heated with coal fires, but which are not really suited to modern central heating. Plants such as the poinsettia and the African violet (saintpaulia) thrive on temperatures up to 70°F (21°C), but azaleas and green-

1 Some alpines, such as Saxifraga macedonica, can be used as pot plants.
2 The florist's Cineraria is Senecio cruentus, which can be had in white and a wide range of colours. The daisy-like flowers are very showy.
3 The wide colour range of Gloxinias.

house primulas will find such temperatures too high. As a result, they come rapidly into flower and soon drop all their blossoms. At a temperature of 55°F (13°C), an azalea will remain attractive for six or eight weeks; under warmer conditions it will last a far shorter time.

The warm centrally-heated rooms of modern buildings will encourage the introduction of plants like the poinsettia. In the USA the giant cyclamen is now unknown as a flowering pot plant for home decoration, as it will not survive at the high temperatures that are considered necessary in that country. It is considerably less popular in Britain than it used to be. This may be not only because many people now keep their rooms at too high a temperature to suit these plants, it may also be because people have realised that it is extremely difficult to keep in good condition in the home.

Commercial growers found that it was necessary to devote a greenhouse entirely to the cultivation of cyclamen, and amateurs have found that they do not mix easily with other greenhouse plants, so that altogether they are not the best of plants for room decoration. Apart from this, they are showy and can be produced fairly cheaply and will, continue in flower over a long period in the right conditions. Given a temperature of around 55–60°F (13–16°C) and good lighting they can be kept going in rooms, but they are very liable to become drawn and floppy.

Watering, which is a problem with pot plants, is rather more critical with cyclamen. It is a truism to say that more plants are drowned than are lost any other way. Indeed there is now an opposite tendency and plants are sometimes kept too dry. Plants that are in flower or shortly to come into flower are thirsty plants and will usually take ample supplies of water. It is not a good idea to turn on the cold tap and use the water immediately. This water is generally far too cold and its application will lower the soil temperature and either

1 All the Begonias are colourful pot plants. Begonia 'Cleopatra' has numerous small pale pink flowers.
2 Primula obconica, a popular flowering plant, is useful for conservatory and room decoration.
3 Solanum capsicastrum, the Winter Cherry, bears large red berries in winter, and is useful for Christmas decoration. The berries last well.

check root growth or possibly kill the roots altogether.

Generally, the water should be at the same temperature as the atmosphere. To achieve this you can either store your water for 24 hours in the room in which your plants are, or, more simply, you can add some hot water to the cold to get the correct tepid temperature. If you have facilities for collecting rain water, which is much appreciated by the plants, then you will want to store it until it reaches

room temperature. The frequency with which you water will depend on a large number of factors. There is more evaporation from clay pots than from the plastic ones, so that clay pots will require water more frequently. However, if the pots have been plunged in some absorbent medium, there will be less evaporation from clay pots. Many plants, among which begonias and fuchsias are more prone to this, are liable to drop their buds if either the soil or the atmosphere is too dry.

Excessive wetness can also cause bud-drop. When plants are watered, this should be done thoroughly. The whole of the space between the soil level and the rim of the pot should be filled with water, which should percolate through the soil and moisten it all throughout. If the water appears to run straight through and out at the end, it is a sign that the soil has become too dry. In this case place the pot in a bucket of water, the level of the water being above the rim of the pot, and leave it there until bubbles cease to rise to the surface. Then remove the pot and firm the soil around the edge of the pot with your thumbs. After that you should be able to resume normal watering.

Composts without soil and mixtures

---

1 Azaleas of various kinds, apart from the so-called 'Indian Azaleas' sold by florists, make excellent, free-flowering pot plants. They will flower over a longer period if the temperature in the room is not too high.
2 Clivia miniata, the Kafir Lily, is a South African bulbous plant, good for the greenhouse or home.

1 The silky 'tassels' of Celosia plumosa, the Prince of Wales's Feather, are long lasting.
2 Begonias of the semperflorens type are available in various colours.

that are based mainly on peat, such as are used for azaleas and winter heathers, will dry out more rapidly than composts that contain a large quantity of loam. The fact that the surface of the soil appears dry does not necessarily indicate that the whole soil ball has dried out. However, when you see the surface looking dry it is usually safe to water the following day. Azaleas have a sort of tide mark on their trunk, which gives you an indication of their requirements. It should be about half an inch above the level of the soil. If it is higher, the plant has had too much water, if it is lower it requires watering. Flowering pot plants are best purchased when they still have plenty of well-developed flower buds. If the plant is in full flower it will not last for long, while if the buds are too immature, they may not recover from the shock of changing atmospheres. It is not unusual for cyclamen to be offered for sale with two or three flowers out and a number of barely-developed flower buds. It is very unlikely that these buds will develop under room conditions and such plants should be avoided.

The majority of pot plants that are offered commercially thrive best in a temperature ranging from 55–60 F (13–16 C) and cannot be expected to persist for so long with higher temperatures. One or two such as cyclamen and cinerarias may even dislike very high temperatures and collapse. Plants such as azaleas and greenhouse primulas will tolerate high temperatures, but will flower and fade at great speed. During the winter months all plants should be given as much light as possible.

The following is a list of principal pot plants for room use arranged according to the months in which they flower: *January* Begonia 'Gloire de Lorraine'; *B. manicata; Rhododendron simsii (Azalea indica);* cultivars, *Erica gracilis* and *hyemalis;* Poinsettia; Cyclamen. *February* Indian azaleas; Cyclamen; *Primula malacoides*, Hippeastrum. *March* Indian Azaleas; Cyclamen; *Primula malacoides* and *P. obconica;* Cineraria. *April* In addition to the March plants, schizanthus, calceolaria, *Primula sinensis;* and the Easter Cactus, *Rhipsalidopsis rosea.* *May* Schizanthus, Calceolaria, forced hydrangea, regal pelargoniums, *Boronia megastigma. June* Regal pelargoniums, zonal pelargoniums, hydrangea, fuchsias, *Rochea coccinea. July* fuchsia, *Exacum affine*, Alonsoa, gloxinia, begonia; *Campanula isophylla. August* As July with browallia, achimenes; torenia *September* As August, chrysanthemum. *October* chrysanthemum, gesneria, × Smithiantha, cyclamen. *November* Chrysanthemum, begonias, poinsettia, cyclamen, *Capsicum annuum. December* Poinsettia, Begonia 'Gloire de Lorraine', *Solanum capsicastrum*, azaleas, cyclamen, Christmas cactus, *Capsicum annuum, Erica gracilis* and *E. hyemalis.*

Saintpaulias can be purchased in flower in most months of the year.

Propagation

The propagation of most pot plants is not difficult, especially if a greenhouse is available, although much can be done on a window-sill. The basic equipment is a supply of pots, seed-boxes and seed, potting and cutting composts (see pages 32 to 34).

Seeds In the greenhouse seeds are generally sown in receptacles of various kinds. For some specialised sowings and for limited sowings clay seed pans are often used. Pots, now offered in a wide range of materials, are most useful for those plants which resent the root disturbances which occurs when pricking out seedlings raised in boxes or pans. New clay receptacles should be soaked in water for several hours before being used. Used wooden boxes should be washed in mild disinfectant.

Where the sowing is made in a seed box some compost is placed in the box, firmed with the fingers and levelled with a ruler or strip of wood to within a half inch of the top. A wooden presser simplifies this task. The compost is then watered thoroughly, using a very fine

1 When seed is sown in pots the surface can be firmed with another pot.
2 Seed sown in drills in a seed box.
3 Seed sown broadcast in a seed box.

rose on the can, and set aside to drain. The seed is then sown thinly on the moist surface. Large seeds are spaced at an inch or so apart. The seeds are normally covered after sowing by sieving a ¼-inch layer of the compost over them. Sowings in pans and pots are carried out in the same way.

The containers are then covered with a sheet of glass on which a sheet of brown paper is laid. The glass prevents evaporation and lessens the need for further watering. The paper provides dark conditions.

Cuttings There are various different types of cuttings which are widely used for propagation purposes. The parts of a plant used may consist of young, green stem-growths, semi-ripe wood, hardwood, single leaves, buds and roots. Stem cuttings may be taken about 3–4 inches in length of half-ripe shoots in July or August of such plants as cistus, hydrangeas, hebes and the like. Some, such as those of camellias, may have a heel of the old wood attached ('heel' cuttings), though most cuttings are prepared by trimming them just below a node or joint ('nodal' cuttings) with a sharp knife or razor blade. The cuttings should be inserted to about a quarter of their depth in pots of moist sandy soil, or John Innes cutting compost, or a mixture of sand and peat, or in a sandy propagating bed in a cold frame. Such cuttings should be shaded from the direct sunlight and be lightly sprayed over with tepid water each morning until roots have formed. Any cuttings that show signs of damping off should be removed.

In the spring young shoots may be taken from the base of such plants as chrysanthemums and dahlias which have been brought into early growth in a warm greenhouse or frame. These are known as soft stem cuttings and after they have been prepared by trimming them cleanly below a node or joint and removing the lower leaves, they should be inserted to a quarter of their depth in moist, sandy soil in a propagating frame with a temperature of about 55°F (13°C).

---

1 Stem cuttings can be taken of many plants, including Dracaena. The stem is cut into short sections with sharp secateurs, each section bearing a bud or eye.
2 The sections are pushed firmly into peaty compost, singly in small pots. Let the top of the stem protrude a little above the compost.
3 The new young plant which results.

4 Stem cuttings made of Dieffenbachia result in a new plant. Three stages of growth are shown.

5 The right conditions for encouraging root formation can be simulated indoors by putting a plastic bag over the top of the pot and securing it with a rubber band, around the pot itself. Zebrina pendula cuttings, rooted indoors in this way. Many plants can be rooted on a windowsill in this way.

Leaf cuttings Healthy, well-developed leaves of numerous plants provide a useful means of propagation. Those that root particularly easily by this means include various begonias, such as *Begonia rex,* gloxinias, saintpaulias, streptocarpus, and some ferns, both tender and hardy. After removing a leaf from the parent plant make a few light incisions with a sharp knife across the veins on the underside and then lay the leaf on the surface of moist compost, consisting of peat and sharp sand. Peg the leaf down gently; hairpins are useful for this purpose. Leaf cuttings should be shaded from direct sunlight and have a reasonably warm and moist atmosphere. Begonia leaves, among others, will produce roots quite quickly, even when just placed in a saucer of water, but the difficulty is that the roots are so tender that potting on the young plantlets is

1 Soft cuttings can be taken of such plants as Coleus. Non-flowering shoots should be chosen and cut off cleanly with a sharp knife.
2 The cuttings are trimmed just below a joint, and the lower leaves removed.
3 The cuttings are dibbled in, being set around the edge of the pot in a sandy compost.

quite a problem.

Camellias are frequently increased by means of leaf-bud cuttings, which are similar, except that the leaf is taken from the current year's growth, complete with a plump, dormant bud with a small piece of stem wood attached. Such leaves are inserted in sharp, moist sand in pots or in a propagating frame in March in gentle heat. With the aid of mist propagation it is possible to deal with much larger numbers of cuttings over a longer period and the percentage that root is usually greater.

Rooting cuttings in polythene film An

1 Leaf cuttings can be made of several plants, including Streptocarpus. Slits are made in the leaf veins, and the leaf is pegged down on to sandy compost.
2 New plants appear at the incisions.
3 The new plants are seen here complete.
4 The new individual plant.
5 Leaf cuttings from a Saintpaulia.
6 Use a small pot and sandy compost.
7 and 8 The cutting and new shoots.
9 and 10 Begonia rex leaves, if cut up and laid flat on sand, will root.
11 and 12 Trimmed sections of Sansevieria leaves grow round the edge of a pot.

Variegated plants are propagated vegetatively to keep their leaf colour.
1 A leaf-bud cutting is made in a variegated form of Ficus by making sharp cuts between the nodes, and leaving a bud at the base.
2 A small section of stem is left.
3 Cuttings are potted up singly and a tray of compost is kept in readiness to provide material for firming.
4 Rubber bands are used to hold the leaves together.
5 Taking a leaf and bud at the base to make a leaf-bud cutting of a Camellia.
6 A single leaf-bud cutting.
7 The cutting potted up singly.
8 Placing hard-wood cuttings on sphagnum moss for rooting in polythene
9 The polythene is folded over.
10 It is then rolled over from one end to the other.
11 Rubber bands are used to make a neat parcel of the cuttings.

---

interesting way of rooting hard-wood or semi-hard-wood cuttings without inserting them in the normal rooting compost, is to use polythene film. The cuttings are prepared in the normal manner and a piece of film about 8–9 inches wide and, say, $1\frac{1}{2}$ feet long, is placed on the propagating bench. On one half of this, along the length, is placed a layer of damp sphagnum moss. The cuttings are then placed on this (their bases may first be dipped in hormone rooting powder if desired) about $\frac{1}{2}$–1 inch apart, their tops projecting over the edge of the polythene strip. The lower half of the strip of film is then folded up over the moss and the cuttings. Then, starting at one end and working towards the other, the strip of film with the moss and cuttings is rolled up tightly and tied top and bottom with raffia or fillis. Roots should eventually form and these will be visible through the clear polythene. When all or most of the cuttings have rooted the roll can be untied and the cuttings potted up or planted out, taking care not to break the brittle young roots.

The advantages of this method are that once the roll has been tied up no further watering is needed as moisture will not evaporate through the film (the roll should, however, be kept out of direct sunshine, on the greenhouse shelf or bench, or even on a window-sill), and that a number of cuttings can be rooted in a quite small space.

Aerial layering or air layering This method is useful for such house plants and greenhouse plants as the india-rubber plant, *Ficus elastica* and its varieties, the related *Ficus lyrata*, and dieffenbachias and dracaenas. It is a particularly useful method of propagating an india-rubber plant once it has become leggy, i.e. when the lower leaves have fallen, leaving an ugly bare stem. A complete ring is cut around the shoot with a sharp knife about 9 inches from the top, and a generous handful of damp sphagnum moss is placed all around the wound, after any leaves on this part of the stem have been removed, but not those at the tip. The moss is kept in place with a piece of polythene; black polythene used in a double layer is best for this purpose. This is then carefully tied top and bottom to form a neat little bundle. Healthy young pencil-thick shoots of ripe wood of the previous year's growth should be chosen for air layering, not old, hard wood. April is considered the best month to do this work, but in a late season May or early June may prove satisfactory. When roots are to be seen in the moss, the layer should be severed from the parent and the young plant very carefully potted in an appropriate soil compost. This is tricky, for at this stage the roots are tender and easily broken, also the period during which the plants are making further roots into the soil compost can be a difficult

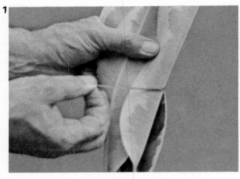

Air layering of Ficus elastica.
1 The leaves at the top are tied together to keep them out of the way.
2 The leaves are removed on a section of stem below the growing shoots.
3 Either a slanting cut is made behind a bud, as shown, or a complete ring of bark can be removed around the shoot in preparation.
4 Damp sphagnum moss is the rooting medium used for air layering.
5 Moss is tied around the cut part of the stem.
6 The moss is covered and protected with plastic film.
7 The top leaves are released, and the work is completed.

one, so that they will require careful attention and nursing until they have become adjusted to their new conditions. Careful shading from direct sun and daily spraying with tepid water will assist them at this period, either in a cold frame or cold greenhouse. The time for an air layer to make roots varies considerably with the type of plant, in some instances it may be many months.